SUBJECT ORIENTED PROGRAMMING

GODWIN ANI

authorHOUSE

AuthorHouse™ UK
1663 Liberty Drive
Bloomington, IN 47403 USA
www.authorhouse.co.uk
Phone: UK TFN: 0800 0148641 (Toll Free inside the UK)
UK Local: (02) 0369 56322 (+44 20 3695 6322 from outside the UK)

© 2024 Godwin ANI. All rights reserved.

No part of this book may be reproduced, stored in a retrieval system, or transmitted by any means without the written permission of the author.

Published by AuthorHouse 03/11/2024

ISBN: 979-8-8230-8673-8 (sc)
ISBN: 979-8-8230-8671-4 (hc)
ISBN: 979-8-8230-8672-1 (e)

Library of Congress Control Number: 2024905627

Print information available on the last page.

Any people depicted in stock imagery provided by Getty Images are models, and such images are being used for illustrative purposes only.
Certain stock imagery © Getty Images.

This book is printed on acid-free paper.

Because of the dynamic nature of the Internet, any web addresses or links contained in this book may have changed since publication and may no longer be valid. The views expressed in this work are solely those of the author and do not necessarily reflect the views of the publisher, and the publisher hereby disclaims any responsibility for them.

CONTENTS

Chapter 1 Introduction 1
The Need for SOP 2
What Is SOP? 5
SOD 6

Chapter 2 Fundamentals 9
Subject and Object 10
Syntax 11
Dimension 11
Cohesion 12
Coupling 13
Composability 13
Programming 16
Encapsulation 18
Composition and Inheritance 19
Exploration 20

Idea Density 20
Common Identity 20
Generalisation 21
Generalisation and Polymorphism 21
Factorisation 23

Chapter 3   Basic Principle Of Sop 25
Relevance of SOP to OOP Languages 31

Chapter 4   Design with SOP 33
Designing, Building, and Using a House 33
Designing a House 35
Building a House 37
Using the House 38
Designing Data Structure 53
Array 53

Chapter 5   Beyond Basic 55
SOD 55
SOP Syntax and Semantic with Example 56
Designing Algorithm 68
Sorting 69
Summary 72

CHAPTER 1

INTRODUCTION

A software application runs in orchestrations of components, modules, libraries, and nowadays micro-services to achieve its functionality. How these components are written impacts how the functionality is achieved, maintained, and improved. These components, modules, libraries, or application programming interfaces (APIs) add to the complexity of the orchestration and the overall software functionality. Complexity in software applications is incredibly difficult to manage. Subject-oriented programming (SOP) is the way to manage or deal with complexity in software applications in ways attempted by object-oriented programming. A typical approach to developing software applications is object-oriented programming (OOP), which is predisposed to complexity and most often more than expected software code in the context of functionality. For example, with typical OOP, the command *Design a house* would create an object (house) and many sub-objects that would extend the house as blueprint along the lines of a bungalow, castle, mansion, and so forth. Meanwhile, the SOP approach cuts through to the function of the house: House shelters occupant (chapters 3 and 4). Much complexity is cut out of the code by starting this way. SOP is the new and better way of programming objects. Most important, this book exemplifies the possibilities are available by using SOP.

The Need for SOP

This book enhances the knowledge of people with above-average OOP knowledge and starts everybody else very simply with SOP without assumptions. Therefore, OOP is not actually explained here as it is not a prerequisite for SOP. SOP means designing and programming software as if it were objects that can be felt or appreciated as abstract or physical objects. OOP tried to do this but couldn't achieve it. An example use of SOP will make this clear. However, a good number of mainstream programming languages like Java, C++, and others implemented OOP. Designing and implementing software in any of these languages usually require but do not necessarily need the use of universal modelling language (UML) and sophisticated integrated development environments (IDEs), respectively.

Many principles go with designing and implementing software in OOP languages, such as single responsibility, open-close, Liskov substitution, interface segregation and dependency inversion principles, SOLID principle, and test-driven development (my personal favourite). But it is still very difficult to write maintainable, readable, and decent-performing code. Data structures and algorithms that are fit for purpose are still incredibly hard to develop in OOP-based languages. For some of us in the high-performance computing space, we find the standard library in these languages nothing to write home about for use where performance really matters. I used to struggle with designing and writing code that performs well in Java. It goes without saying that software code must be readable, maintainable, robust, error-free, and run really fast (or let's say fast enough for the given software application). I mean, software should continue to be fast, reliable, and maintainable as new features are added to it. Latency and throughput should still be predictably guaranteed even as loads increase. Concurrent and distributed systems should be easy to program with an error-free guarantee. The development of SOP is the way to ensure that software can have these qualities.

I think the software industry has battled long and hard with OOP, so much so that it's time the industry designed and programmed with objects the proper way by using SOP. The industry has gone through one bad design decision after another in the history of programming languages, from the use of null to layers of abstractions in operating systems, virtual machines, cloud platforms, and containers that don't align well with memory. I am sure many appreciate that these layers of abstractions have contributed immensely to the unnecessary challenges software engineers face in writing software that can relate well with memory and cache subsystems and with expected good performance. I think that aligning with the hardware may not be a bad idea as these layers of abstractions are sometimes unnecessary. Designing software with SOP will certainly help to be as aligned with the computer as possible and yet at as high level as required. Admittedly, tackling these problems is not easy and has not been easy. But they are, simply put, design problems, and SOP is the answer to them.

Strictly functional programming languages are not exempt from the issues highlighted previously, even as they claim to be based on mathematics. Provocative, right? They clearly use virtual machines too. But besides that, functional programming also has its peculiar issues. The issues are exactly why pure functional programming cannot do input/output (I/O) in the context of computer system, which is, ironically, the essence of computing.

SOP nicely solves all the design issues we currently have in software. Nevertheless, there is a bit of a learning curve for SOP. It can take a bit of time and practice to master it. But software consultants, architects, engineers, and developers are familiar with the challenge of learning new ways of doing things. Therefore, there is plenty of space in SOP for highly skilled software professionals to separate themselves. You will get a good sense of the design depth SOP can help one achieve with precision by the time you finish reading this book. Personally, I enjoy developing software for low latency and high throughput without compromising the qualities—maintainability, readability, robustness, and so on—of a well-engineered software application. Therefore, it is important to me that SOP ensures all of these qualities during design. This is why the SOP design (SOPD) process involves manipulating actual SOP code (SOPC), and the product of the process is SOPD. This has nothing to do with drawing diagrams. All one needs for SOPD is a computer with a basic notepad on it, as will be demonstrated later.

However, the use of an IDE for SOPD is in order as SOPC is similar to, if not the same as OOP language such as Java.

Having warmed up a bit, let's discuss what SOP is.

What Is SOP?

SOP is the way of modelling and programming with objects using English grammar and semantics as the basis. It involves considering the meanings of objects being programmed in a sentence. Therefore, understanding basic English grammar is key to doing SOP. The subject in SOP is the analogy for the grammatical subject of a sentence. Subject orientation emphasises the use of a simple English sentence to express exactly the problem of interest. A simple sentence has one subject, one verb, and at least one object. One verb explains that the sentence predicate consists of just one verb or action. The subject is an object, but it is a special object because of its role and position in a sentence. The subject is normally the first object in a sentence, which performs the action or acts on the other objects of the sentence.

An English sentence is the best way of expressing the relationship between objects. The relationship between objects is what captures the problem being expressed. It is also the context of interest for the objects within which they are programmed. For example, the sentence, "House shelters occupant," expresses House as the subject, shelters as the action, and Occupant as the object. The relationship between the two objects is *shelters*, which is also the action. The sentence can be represented in SOP as:

```
House.shelters(Occupant).
```

House, as the subject of the sentence, performs the action *shelters* on the object Occupant. This is the natural language of real objects expressed in an English sentence. Simply put, House and Occupant are the real objects being programmed to solve the problem *shelters*.

This is named SOP because this method of programming with objects is inspired by English grammar. Every well-constructed English sentence consists of a subject, a predicate, and objects. These basic concepts of grammar can be explained by the subject directs an action (predicate) to the object.

SOD

SOD is simply designing and programming objects, both real and abstract, from the first principle. SOD uses SOP, exploration, and idea density as core principles of design. Chapters 4 and 5 demonstrate SOD.

The core principles of SOD are exploration and idea density. Exploration is a process of exploring a given problem or solution in a concise and representative form that supports clear reasoning. Idea density is a principle that underpins the use of the English language and mathematics to design software. Idea density captures the fusion between English and mathematics for problem-solving. Other principles and concepts of SOD are generalisation, basic/

common identity, and factorisation. Chapter 2 illustrates these principles.

In SOD, generalisation can be done at both high and low levels of abstraction. The most important one is the low (or detailed) level of generalisation, which is where basic/common identity principles come in as the principles that guide generalisation at the lowest level of detail. Factorisation is exactly what it means in mathematics, only it can also be applied to both English sentences and words as well as numbers. SOD can be used to design software to any level of detail—from high-level design through actual classes/components to data structures and algorithms.

To demonstrate SOP, the following examples have been selected to show the essence of SOD and SOP. And in chapter 4, we discuss designing, building, and using a house, and designing data structure. In chapter 5, we discuss SOP syntax and semantics with examples, designing algorithm, and sorting.

CHAPTER 2

FUNDAMENTALS

Every subject must have a single responsibility. This single responsibility defines the reason or rationale for using or creating the subject given context. This means that if single responsibility does not justify the essence of a subject, there is a need to ensure this happens. This can be done by considering carefully what the subject should be responsible for given context. For example, let's consider different sentences that involve a house and an occupant.

- House shelters occupant.
- House accommodates occupant.

There are not many sentences that can be made in the context of what house can do for or to the occupant with the objects House and Occupant. This reduces the number of responsibilities to consider to the fewest number of sentences. That makes it quite hard to miss the responsibility that is required or needed. On the other hand, it makes it easy to cut out all the irrelevant responsibilities to focus on the responsibilities of interest, such as the two above.

Let's consider the two responsibilities in the two sentences. The first responsibility relates to protection; the second relates to space. This simple exploration of the two responsibilities has made it easier to understand the responsibility of interest. If protection is chosen, this code

```
House.shelters(Occupant)
```

suffices as the responsibility of interest. Otherwise, it is space, which means

```
House.accommodates(Occupant)
```

Accommodation is dealt with elaborately in Chapters 4 and shelters is as well dealt with in chapter 5. Fundamentally, subject and object programming is done with the verb in context such as shelters or accommodates in between house and occupant.

Subject and Object

An instantiated class is an object or a subject. An instantiated class is basically class that is loaded in memory and interacting or ready to interact with other objects or subjects, depending on which is sending or receiving a message or more appropriately action. For example, *Architect designs a House*, can be represented in SOP as:

```
Architect.designs(House)
```

In the code above Architect is the subject, and House is the object as the architect sends the message or acts—*designs*—on house.

Syntax

SOP syntax is subject syntax represented as

```
Subject.method(Object)
```

This is a recognisable basic subject/object orientation syntax. As Architect is subject, *designs* is method, and House is object. The argument for this method is:

```
Architect.designs(House)
```

Next is the SOP-fluent syntax, where input and output to/from method are represented.

```
Subject.method(Object):OutputObject
Architect.designs(House):Project
```

Project is the output of the *designs* method, which can form the input of another method in composable-fluent API. SOP syntax and semantics are explained in detail in chapter 5.

Dimension

Dimension is the context and scope of a given problem of interest represented by Subject Oriented Programme as:

```
Subject.methods(Object):OutputObject
```

Or

```
Subject.ObjectArgument
Subject.methods(Object):OutputObject
Object.methods1(ObjectArgument):OutputObject
```

This is the framework of subject oriented design with which every object programming problem is analysed and solved as used in this book. The framework reads: subject methods object for OutputObject by object methods1 ObjectArgument for OutputObject.

Analogically, dimension is a unit that can be tested as in Test Driven Development (TDD) unit testing. Dimension expresses practical illustration of subject orientation that can be used to solve problems of subject and object programming. Dimension illuminates the crux of subject and object programming in a representative concise fashion. Let's have a look at core concepts of subject and object programming such as cohesion, coupling, etc. in the context of dimension.

Cohesion

Cohesion is the degree to which subject or object is focused for understandability and readability expressed in subject's methods. In the context of dimension, subject is highly cohesive if the methods almost verbalises the subject. For example, in Java programming language, there's a class called Runnable with only one method called 'run'. Runnable 'run' is highly cohesive. House 'shelters' is highly cohesive. However, dimension provides the structure with which cohesion is considered. Therefore, with dimension it will always be clear how cohesive the method is with the subject. High cohesion is encouraged as opposed to otherwise. High cohesion means reusability and modularity of the dimension. Generally, every method should express the associated subject or object to the highest degree possible that way the functionality there in is focused and manageable.

Coupling

Coupling is the extent or degree to which subject and object in a dimension are connected. Coupling needs to be as low as possible for clear separation of concerns and responsibility assignment. For example, in the dimension framework or structure, the subject passes ObjectArgument to the Object via method1. This is the way to keep dependency between subject and object as low as possible, by argument passing from subject to object and not otherwise or the other way round.

Composability

Composability is the common identity between the Subject and Object in a dimension or the common identity between dimensions in the context of architecture or integration. For example the composability of standard subject oriented programming syntax:

```
Subject.methods(Object):OutputObject
```

is ObjectArgument as shown in two different ways below. There are composability by Subject member and composability by Object argument.

Default option: Subject Member

```
Subject.ObjectArgument
Subject.methods(Object):OutputObject
Object.methods1(ObjectArgument):OutputObject
```

Technically, this means that new Subject has to be instantiated each time the 'methods' is invoked. This is the overhead inherent in subject member option, which is performance overhead on programming language runtime such as Java, Scala, Kotlin, etc.

Alternative option: Object Argument

```
Subject.methods(Object, ObjectArgument):OutputObject
Object.methods1(ObjectArgument):OutputObject
```

Object argument option does not have the sort of overhead that Subject member option does with new subject instantiation on each invocation of 'methods'. However, Object Argument changes the semantics of the dimension or unit as in Subject 'methods' Object with ObjectArgument for OutputObject instead of just Subject 'methods' Object for OutputObject.

Let's assess a simple example of composability: House 'shelters' Person as Condition, with Person 'controls' Condition as condition represented as:

```
House.shelters(Person): Condition
Person.controls(Condition): Condition
```

Composing with Subject Member option, House has to have condition as member for house to compose with person, hence the following code:

```
House.Condition
House.shelters(Person): Condition
Person.controls(Condition): Condition
```

Alternatively, composing with Object Argument option, 'shelters' has to have another argument Condition next Person Object as illustrated below:

```
House.shelters(Person, Condition): Condition
Person.controls(Condition): Condition
```

In the context of architecture or integration composability options is normally, object argument option, and hybrid of subject member and object argument, which characterises fluent composable API. In principle the output of the previous feeds in to the input of the next as long as there are common identities between the dimensions or units.

```
Alternative option 2: Output Object
Subject0.methods0(Object0):Object
Subject1.methods1(Object):OutputObject
```

The composition is as follows:

```
Subject0.methods0(Object0):Subject1.
methods1(Object):OutputObject
```

Composability provides clarity on the trade off between the performance and semantics overheads in Subject Oriented Programming as illustrated in this section.

Programming

Every subject oriented programming code is normally targeted for implementation in modern programming language such as Java. Translation of subject oriented programming code to Java for example is about one to one mapping most often than not. The value of SOP code is tremendous as it is also for one to one mapping to modern programming languages such as Java, Python, C++, etc. Java is the programming language of choice for this book. Programming subject oriented programming code is like programming with Java's one method interfaces such as Runnable, Callable, Functions and Lambdas. Having said that, translating typical SOP dimension to Java is as follows.

SOP Dimension:

```
Subject.ObjectArgument
Subject.methods(Object):OutputObject
Object.methods(ObjectArgument):OutputObject
```

Java Translation:

```
public class Subject {
   private final ObjectArgument objectArgument;

   public Subject(ObjectArgument objectArgument) {
      this.objectArgument = objectArgument;
   }
```

```
public OutputObject methods(Object object) {
    return object.methods(objectArgument);
}

}
```

For example the below actual code:

```
House.Condition
House.shelters(Person): Service
Person.controls(Condition):Void
```

translates to Java equivalent with indented Person.controls(Condition):Void translated as Java statement in the following java code:

```
public class House {
    private final InOut[] in = new InOut[]{InOut.VACANT,InOut.SIT};
    private final InOut[] out = new InOut[]{InOut.STAND,InOut.VACANT};
    private final Condition condition;

    public House(Condition condition) {
        this.condition = condition;
    }

    public Service shelters(Person person) {
        person.controls(condition);
        return new Service(in, out);
    }

}
```

```
public enum InOut {
    SIT, STAND, VACANT;
}

public class Person {
    private Condition condition;

    public Person(Condition condition) {
        this.condition = condition;
    }

    public void controls(Condition condition) {
        this.condition = condition.merge(this.condition);
    }
}
```

Encapsulation

Encapsulation is concept well known in the computer science, software industry, and information technology at large. Encapsulation is also poorly understood and seldom properly implemented in the software industry. In the context of the software industry, encapsulation means that every object or class must have a single method and a single behaviour. This seems too stringent, but that's how real-world objects behave. For example, a house shelters, full stop. Some may argue that's the real world, but in software, it is different. In software, if there are multiple methods in a class or object, encapsulation is already broken, whether through private or public means.

However, if you have more than one method in a class, object, or subject, it means that you are starting to program and still need exploration, idea density, common identity and dimension framework to get to encapsulation. In the software industry today, there are issues and compounded problems with APIs, modularisation, library, micro-services, and architecture in general because of fundamental encapsulation problems at the level of classes. SOP principles ensure encapsulation in a way that takes software development in the right direction for good. SOP encapsulation localises a behaviour in a single method for a given class, subject, or object.

Pure encapsulation may seem futuristic in object programming but Programming language like Java is already headed in this direction. For example, Java has one-method interfaces/classes in Runnable and run, Callable and call, Functional interfaces/Classes, and lambda. With these, programming with design patterns has become idiomatic in Java among other benefits such as cohesive APIs.

Composition and Inheritance

Inheritance is about reuse of behaviour. Inheritance as an SOP concept enables a subject to inherit behaviour from another object. Since inheritance is about using already existing behaviour of another object, the way to inherit behaviour is by composition. In SOP, there is no other way of doing this but by composition. It is also important to note that composition is the best way of inheriting behaviour in SOP as well as OOP. See composability section for details.

Exploration

Exploration is a process of exploring a given problem or solution in a concise and representative form that supports clear reasoning. However, in the context of SOP exploration is the process of ordering the use of idea density, common identity, factorisation, and generalisation in a specific view or abstraction.

Idea Density

Idea density is a principle that underpins the use of English language grammar, words, sentences, and meaning to guide programming computers, hence SOP. This is used very extensively in the subsequent chapters. Therefore, there is no need for any example here. Idea density also underpins the use of English language and mathematics together to guide programming computers. For example, this enables the factorisation of common identities in some exploration contexts, as will be seen in subsequent chapters. Idea density at this point sounds very dense, but its use in the design of a house in the subsequent chapter exemplifies its power very simply for a layperson. Idea density captures the fusion between the English language and mathematics for problem-solving using SOP.

Common Identity

Common identity is a concept that identifies common subject, object, behaviour, or method between subjects in consideration. Common identity enables generalisation and factorisation at different levels of abstraction. For example, the common identity of an array of integers is the indexes of the integers in the array. This is illustrated in chapter 5, and there are many examples of the use of common identity in this and subsequent chapters.

Generalisation

Generalisation is a concept used in SOP in a different way from its normal use, which is at high level of abstraction. In SOP, generalisation can be done at both high and low levels of abstraction. The most important one is the low- (or detailed-) level generalisation, which is where basic/common identity principles guide generalisation at the lowest level of detail.

Generalisation and Polymorphism

According to Wikipedia: "Polymorphism is the use of a single symbol to represent multiple different types". In the context of SOP, polymorphism is about generalisation of subjects and objects as illustrated with SOP dimension framework in this section.

Let's design animal or pet's sounds in SOP.

Cat can make Sound as well as Dog, but they make different sounds.

```
Cat.makes(Sound)
Dog.makes(Sound)
```

What is the common identity between Cat and Dog in the context of Sound? It is mouth. So Cat and Dog can be generalised to Mouth.

```
Mouth.makes(Sound)
```

This is a typical specific generalisation of the subject at low level of abstraction. However, for object generalisation, pet is a polymorphic concept with types: cat, dog, etc., but importantly as always, pet can vocalise sound with its mouth. Using SOP dimension framework, the generalisation of subject and object is as follows:

```
Mouth.vocalises(Pet):Sound
Pet.speaks():Sound
```

Similarly for cat and dog

```
Mouth.vocalises(Cat):Sound
Cat.speaks():Sound

Mouth.vocalises(Dog):Sound
Dog.speaks():Sound
```

The generalisation reads: Mouth vocalises pet for sound and pet speaks for sound. Notice the use of vocalises instead of makes as more focused word contextually. Not only does this read really well, it is implementable program - polymorphism. The low level generalisation is completed by applying the SOP dimension framework. Again, these can be translated to actual Java code for example, in similar way as in programming section. Translating SOP code to Java programming language is as follows:

```
public class Mouth {
    public Sound vocalises(Pet pet) {
        return pet.speaks();
    }
}
```

Subject Oriented Programming

```
public sealed class Pet permits Cat, Dog {
   public Sound speaks() {
      return new Sound("Pet");
   }
}

public final class Cat extends Pet {
   public Sound speaks() {
      return new Sound("Cat");
   }
}

public final class Dog extends Pet {
   public Sound speaks() {
      return new Sound("Dog");
   }
}
```

Though translation of SOPC to programming language of choice, which is Java is not the main focus of this book (see chapter 4 and 5 for more examples), but fundamental use of SOP is the focus.

Factorisation

Factorisation is exactly what it means in mathematics, only it can also be applied to English sentences and words as well as numbers. Furthermore, it applies to subjects, objects, methods, and so on, as well. Factorisation usually features during specific generalisation, which entails factorisation of common identities and then choosing the common identity relevant to the problem being solved. Factorisation is normally used in the context of common identities, as illustrated in generalisation and polymorphism and chapters 4 and 5.

CHAPTER 3
BASIC PRINCIPLE OF SOP

Let's start by asking a simple question: How can I design a house using subject orientation? First, I consider the meaning of a house. What is a house? A house is an object that shelters the occupant. That's it. This is the very basic principle of SOP.

A fundamental rule of SOP is that every subject must have *exactly* one behaviour. I mean one method per subject or object. Some might argue that one method per subject or object is too granular and will result in many objects created needlessly for what could have been one object and many methods. Stretching this argument a bit means it is about trade-off between the two extremes as it is done in OOP. In SOP, it is not about trade-offs at all; it is about specific generalisation. For example, in the context of Person, Chair (for sitting) has specific behaviour, which is to *sit*. Some would argue, what if a book is put on the chair as well? The chair should then have at least two methods: *sit* for person and *shelf* for books. In SOP, what changed is the context from person to room. In the context of Room, chair has one behaviour—*support* for person and books, but no longer *sit* or *shelf*. This is specific- context generalisation at a granular level for one method per subject or object.

The only exception for one method per subject or object is an adapter, which has exactly two methods. An adapter is illustrated in chapter 4. In contrast, an OOP object can have a public method and some private methods, or many public methods that are in aggregate a behaviour. A little digression: Aggregate behaviour is obtained in micro-services on the modular level and rightly so. Every object—physical or abstract—has exactly one behaviour or method in a given context. SOP adheres to this basic principle. Therefore, SOP not only strictly preserves that each object has exactly one method, it also helps to achieve exactly one behaviour per method. In SOP, it is one method, one behaviour; that's it. This is encapsulation ensured without question. How to achieve one method, one behaviour per object without creating many needless objects accentuates the power of SOP as the way to design and programme objects. Designs produced by SOP are normally elegant. But in addition, they provide clarity on how well they will perform.

Let's consider the sentence: *A house is an object that shelters the occupant* in the context of grammar and computer. House is the subject of the sentence, *shelters* is the verb or action, and Occupant is the object of the sentence. In essence, the House performs an action on the object (Occupant). This is the natural language of objects in reality, analysed in a sentence. Expressing it in SOPC looks like this:

```
House.shelters(Occupant)
```

This is a recognisable piece of code, especially to those who are nurtured in OOP. This is the type of code manipulated in the SOPD process and the type of code that can be produced as SOPC or design. Programming with objects is about sending messages to objects. This is also the case in the real world. In this context, House sends the message *shelters* to Occupant, the object of the sentence. But Occupant can also be the subject of the sentence. Anecdotally, an occupant is normally a human being who can perform an action on objects. So why not make the occupant the subject of the sentence and house, which is object in real life, the object of the sentence? Let's do just that. Say, "Shelter, occupant, house." That is,

```
shelter(occupant).House
```

The action is now directed to House, and House is the object. But we don't have the subject anymore. We have predicate and object. In other words, we have the message,

```
shelter(Occupant)
```

and House that receives the message. Don't panic, we still have subject, but it is no longer explicit in the code. The subject is the underlying computer system, which actually sends the message to House. Typically, the system sends messages via a computer. This clearly explains why message sending is implementation detail which should be behind the code. Therefore, message sending is outside the scope of this book. Besides, such implementation detail would get in the way of understanding SOP in this context.

There is one more variation in addition to these. Say, "Let occupant be sheltered by house":

```
Occupant.shelteredBy(House).
```

This code gives a sense that the prefix (*sheltered*) of *shelteredBy* is a state of the Occupant, transitioned to by an action from House prior to this state. This defines a state machine as an action from the subject (House) transitioned the object (Occupant) from one state (possibly unsheltered) to another state (*sheltered*). This means that SOP has great potential for designing state machines as well.

Before going further, let's consider the model of SOP core syntax that will be used for the rest of this book.

```
House.shelters(Occupant)
```

In this case, *shelter* is an action as opposed to a message. This clarifies that message sending is done behind the syntax, and the computer system sends messages from one memory cell to another. This is what the syntax promises on the code itself, but the actual sending of a message is done in the implementation of the message-sending model that lies behind the scenes. Evidently, the code House.shelters(Occupant) represents a message as a whole. This means that the subject does not send messages at all. The computer system does. The subject directs action to the object.

The art of directing action to object represents the message being sent by the computer system.

Using the same model of syntax, can House exist without Occupant? Yes, it can. I can have:

```
House.shelters()
```

and it will still be subject orientation. In fact, that is exactly why it is called SOP and not OOP.

Without a subject, nothing directs or performs action on the object, and nothing gets done at all. However, House performs action without directing it to anything in particular. This means that whatever object is interested in the action can receive or take on the action. This means that House provides *shelters* services to objects that are interested in this service. A house can be built to provide shelter to no specific occupant but to any occupant. Therefore, a house without any occupant is still a house and can provide its service when an occupant becomes available. This also implies that the house can have an occupant anytime. In software terms, this is the basis of providing software as services in similar way—hardware as service, platform as service, and so on. These services are not directed to anyone in particular, but whoever pays for them uses the services.

When subject performs an action but did not direct it to any particular object, it does not mean that no object receives it. It means that the objects that will receive it are not explicit in the syntax. It also means that the programmer is free to choose to what object the action will be directed. This is where the all-important side effects can be done: print to the screen, write to file, write to network or to the disk, and all the above. This underpins the basis of software services and/or software as service. It can also be a command. How is it a command? It is a command because *shelter* is an action word.

Using a military analogy to explain this, the nature of command is that the senior officer gives a subordinate a command. The senior officer hopes the subordinate carries out the command unless there is a system that will help to ensure the subordinate carries out the command. In the military, they ensure this by rank and a strict chain of command. Otherwise, no one has any other solution. Due to lack of a similar solution in software allows the opportunity for software service. In software terms, a command can be issued, and the programmer directs the command to a file, for example, where it is written and waits for a paying customer to show up. Once the customer does, the command is revoked from the file and directed to the customer. Of course, there are many ways of doing this in real time and offline. But this is not the focus of this book. Moreover, the difference between service and command in this context is whether there is an 's' at the end of the verb (*shelter*). Therefore, this code,

```
House.shelter()
```

represents a command, but

```
House.shelters()
```

represents service. These things are not new to programmers, by the way.

We are ready to code in SOP. In SOP, it is all about writing and manipulating SOPC. SOP is a universal design language. This means its use can extend beyond software, but it is primarily developed for software design and programming. Therefore, software design is the focus of this book and the examples in chapters 4 and 5.

There are many approaches to doing SOP:

- Top-down approach
- Bottom-up approach
- Exploration approach

Not all of them are covered in this book. However, different approaches will be used for certain case studies in subsequent chapters. The top-down approach to SOP is basically starting from the top layer of abstraction and designing our way down to the finest detail. The reverse is the case with the bottom-up approach: Start from the finest detail one can think of and design up to the top layer of abstraction. The exploration approach combines these two approaches, designing up and down layers of abstractions and generalising as necessary for the given problem. The exploration design approach is used in chapters 4 and 5.

Relevance of SOP to OOP Languages

As mentioned previously, SOP is very relevant to modern programming languages like Java, Python, and C+. SOPC can be programmed directly in Java, for example. SOP is a far better approach to programming objects in any programming language than OOP. SOP is a better paradigm shift from the complexity of

OOP to better software applications. Software developers who adopt SOP can work directly in their favourite programming languages. It is just another way of thinking and programming objects using a computer.

SOP syntax is similar to OOP languages such as Java. The syntax is as important as SOP approach to OP, which is where better API designs, micro-services, modularity, and library composability/reusability are ensured. Syntax and the semantics of SOP are covered in chapter 5.

CHAPTER 4

DESIGN WITH SOP

Design as it is used in the context of SOP is the use of exploration to analyse problems for specific outcomes in the context of computing. This chapter demonstrates how to use SOD—a method used with SOP to design, build, and use a house. The aim is to design or model software problems with SOD. The design model can be coded in any current OOP language. This chapter also demonstrates how to use SOD to design algorithms. Note that some novel principles and ideas are used in this case study. If you read the fundamental principles of SOP (chapter 2), these principles should be familiar.

Designing, Building, and Using a House

This chapter demonstrates how to use SOP to design, build, and use a house. The aim is to design or model software problems with SOP.

The design model can be coded in any current OOP language, such as Java, Python, Scala, Kotlin, or C++. This chapter also demonstrates how to use SOP to model algorithms. As noted earlier, some novel principles and ideas are used in the demonstration, but if you read the basic principles of SOP in chapter 3, they should be a bit familiar. Let's start with the house shelters occupant code that should be familiar by now:

```
House.shelters(Occupant)
```

We know that a house shelters occupants from weather such as rain, snow, wind, and cold. How does a house shelter? A house shelters with a Roof, Walls, and a Floor. The Roof shields the weather from entering the house from the Top. The Walls shield the weather from entering the house from the Sides, and the Floor shields the weather from entering the house from the Ground. The following SOPC represents these:

```
Roof.shields(Top)
Walls.shields(Sides)
Floor.shields(Ground)
```

In addition, these follow:

```
Walls.sizes(Room)
Room.uses(Space)
```

As you can see, we are exploring what a house is and how it does what it does. Because I use the very basic syntax of SOP to design, there is no need for any diagrams or UML. This is the way to design software. I mean the way to design and code any type of system. It is a general-purpose design language that can produce an implementable and a complete design as well as actual working code for the software application being designed. SOP can help software consultants, engineers, developers, and architects to produce elegant software design. SOP can be used to design algorithms and data structures. SOPCs are guaranteed to perform well when implemented in a programming language. This is the way to design and code high-performance computer programs. In theory, any design SOPC is maintainable, readable, robust, ideally error-free, and most important, will perform better than other conventional design methods in the same programming language. SOP is programming language agnostic, which means you can design with SOP and implement the code in any programming language of your choice. It goes without saying that SOPC is guaranteed to perform better than design diagrams produced by tools like Enterprise Architect/UML, and so on. Observably, SOPC looks very much like OOP code in Java, so it can be coded directly into Java as well.

Notice that Walls has two methods in the previous code. It will be resolved as we design. That code presents the beginning of the design process. Now, let's design a house.

Designing a House

In order to design a house in real life, a building architect is usually required. This means that architecture, or style of the house, is very important for design. So, let's start by saying the Architect *designs* the House. The House consists of rooms. The result of the design is a house Design document. Meanwhile, the objective is a software design model of the content of house design. SOPC suffices as the software design model. So, let's achieve this objective.

```
Architect.designs(House): Design
Architecture.styles(House): Design
House.[Interior: Modern, Exterior: Gothic]
Area.dimensions(House)
House.[roomCount, length, width]
Area.dimensions(Room)
Room.[length, width]
```

As shown in the objective, the House design is about styling and dimensioning of the house and rooms. Let's add the capability to choose Bungalow or StoryHouse:

```
Architect.designs(House/Room): Design
Architecture.styles(House): Design
House.[Interior: Modern, Exterior: Gothic]
Area.dimensions(House/Room)
House.[length, width, roomCount]
Room.[length, width]
Structure.shapes(House)
House.[Type: Bungalow/StoryHouse]
```

Observe that (House/Room) means House or Room as objects or arguments. The Design in the

```
Architect.designs(House/Room): Design
```

is the return object from the *designs* method. The square brackets in

```
[Interior: Modern, Exterior: Gothic]
```

represents an array of variables. Interior and Exterior, and Modern and Gothic are their values.

Next,

```
[length, width, roomCount]
```

represents an array of variables without values. Notice also that House and Room are dimensioned in same area subject. It is allowed in SOP as it helps to reduce repetition where necessary. Observe that one can add or remove from the design as required, and the resulting design will still be complete for the given purpose. Note that adding or removing features does not necessarily mean more code. It means that a different but simple complete design can be achieved.

Building a House

In order to build a house, one needs a project that will implement the actual building of the design of the house. The project entails the management of labour, resources, material procurement, and so on. All of these are encapsulated in project. The SOPC from design of the house is also encapsulated in the Design. Building a house involves the actual laying of bricks to erect walls which, in turn, shape rooms, and then furnishing.

```
Project.builds(Design): House
Design.lays(Bricks): Walls
Walls.shape(Rooms): Building
Building.isFurnishedBy(Project): House
```

Nonetheless, this is enough to capture the essence or the purpose of the design. Notice also that I did not have to disambiguate Walls (internal and external) as the purpose of the design does not necessitate doing that just yet. The house has been built in SOP.

Using the House

The use of the house is explored from the perspective of the occupant. The occupant interacts with the house via the rooms and furnishings. Let's say, "Occupant uses the bedroom and makes his way to use the living or sitting room." Applying idea density in the form of dividing bedroom into bed and room seems a bit strange but will surely stay well with the reading in the course of reading this book.

```
Occupant.uses(BedRoom)
```

Splitting BedRoom into Bed and Room allows for the following code:

```
Bed.relaxes(Occupant)
Room.fits(Bed)
```

Let's continue with the living/sitting room.

```
Occupant.uses(LivingRoom)
Occupant.uses(SittingRoom)
```

There is no need for division on this occasion as *sitting* and *room* are separate words normally. They can be used separately, but sitting does not lend itself well to the required subject. Therefore, further

Subject Oriented Programming

exploration of the idea of sitting is necessary. However, this is not far-fetched. In most sitting rooms, one can expect to see a sofa for the idea of sitting to be fulfilled. This same process works for the use of living room. The following code applies:

```
Sofa.relaxes(Occupant)
Room.fits(Sofa)
```

Similarly, the following SOPC should feel intuitive enough for the reader to understand.

```
Occupant.uses(DiningRoom)
DiningTable.dines(Occupant)
Room.fits(DiningTable)

Occupant.uses(Kitchen)
Occupant.uses(Cooker)
Room.fits(Cooker)
Cooker.serves(Occupant)

Occupant.uses(BathRoom)
Bath.washes(Occupant)
Shower.washes(Occupant)
Room.fits(Shower)
Room.fits(Bath)
```

The apparent fluency in coding the DiningRoom, Kitchen, and BathRoom was powered by idea density as it is glossed over gently by the application of division on English words.

Furthermore, notice that the occupant uses the bedroom, sitting room, kitchen, and bathroom quite a lot. This will undoubtedly make the *uses* method complicated and clunky. This seems like a design smell. In order to deal with this, the prefix of each room—BedRoom, SittingRoom, and so on—needs to be removed so that the main furnishing of each room defines the room.

```
Occupant.uses(Room)
Bed.relaxes(Occupant)
Room.fits(Bed)
```

Swapping the subject (Occupant) with the object (Room) automatically removes the responsibilities of the *uses* method from the Occupant and puts them on the Room.

```
Room.uses(Occupant)
Bed.relaxes(Occupant)
Room.fits(Bed)
```

Clearly, a room does not use an occupant, but it does accommodate the occupant. Therefore, let's change *uses* to *accommodates* and Occupant to Person for simplicity

```
Room.accommodates(Person)
Bed.relaxes(Person)
Room.fits(Bed)
```

Accommodates is much more focused than *uses* and is expected to produce a much simpler code. This is idea density resolving responsibility issues in the design. Notice the indentation that effectively says that Bed is a member of Room. The last line of code states the obvious, already captured by the second line. Therefore, the last line should be removed.

```
Room.accommodates(Person)w
Bed.relaxes(Person)
```

Now, let's explore this:

```
Bed.relaxes(Person).
```

When a person relaxes in bed, the bed supports the torso, the bottom, and the legs. Therefore, a bed is a piece of furniture that does all these three things. On the other hand, a sofa also relaxes. The difference is that a sofa typically relaxes the torso and the bottom. However, a La-Z-Boy sofa relaxes the torso, bottom, and the legs. Furthermore, each chair in the dining room relaxes the torso from the bottom. Some chairs can have a footrest, which means that dining chair can also relax the torso, bottom, and legs. For now, let's select the properties that have been identified as common among the three pieces of furniture, which are furniture supports for only the torso and bottom. The following code represents the exploration:

```
Room.accommodates(Person)
Bed.relaxes(Person)
Furniture.supports(Person)
'Basically supports person's torso and bottom'
```

This demonstrates how to generalise at the lowest level of detail. It is clearly enabled by idea density. Notice that Furniture is a member of Bed. This is inheritance by composition— the best way to do inheritance—which is carved out of objects by common identity. In word order, a bed *is* a piece of furniture that can support a person, more specifically, a person's torso and bottom.

I know this looks like Python language indentation. But it makes things clear as long as you keep it to only three levels, which is the standard in SOP. Alternatively, the code can be aligned as follows:

```
Room.accommodates(Person)
Bed.relaxes(Person)
Furniture.supports(Person)
'Basically, supports person's torso and bottom'
```

The two communicate the same things, but the structures are different. This is refactoring, as in test-driving development (TDD). The structure of code was changed without changing the functionality of the code, which is the meaning of refactoring. Observe also that the person using the room and bed do not have any functionality about the room, bed, and furniture. This is a better design. Similar changes are made to bath, sitting, and dining rooms.

```
Room.accommodates(Person)
Sofa.relaxes(Person)
Furniture.supports(Person)
'Basically, supports person's torso and bottom'

Room.accommodates(Person)
Dining.entertains(Person, Food)
Furniture.support(Person)
'Basically, supports person's torso and bottom'

Room.accommodates(Person)
Bath.washes(Person)
Shower.washes(Person)
```

Observe that Dining has two objects, which are basically two arguments: Dining *entertains* Person with Food. When a person enters a room, it is for a purpose. For example, I enter the bedroom to lie in the bed, and I enter the dining room to eat food. As I enter these rooms, I enter as Person. There is no indication what I entered with. I could have entered naked or with item(s), and so on. These details are not mentioned because they are not relevant to the purpose of the design so far.

Let's go back to dining. I can enter the dining room as the master of the house to eat, and a servant enters to serve the food; both are Persons. Therefore, servant and master are all persons, but they capture different things or connotations. So I can reconcile these ideas about entering a room into a denser form, which is still Person. This explains why

```
Room.accommodates(Person)
```

still takes only one object (Person) and suddenly, there is Food in

```
Dining.entertains(Person, Food)
```

as another object or argument. Reiterating, a servant entering the dining room explains the food. At this level of design, there is no need to detail Person any further.

```
Room.accommodates(Person)
Dining.entertains(Person, Food)
Furniture.support(Person)
   'Basically, supports person's torso and bottom'
```

Let's consider the sitting room:

```
Room.accommodates(Person)
Sofa.relaxes(Person)
Furniture.supports(Person)
'Basically, supports person's torso and bottom'
```

Further generalisation focuses on Sofa. A sofa is a type of furniture that provides seats for a person. But one can put a book on a sofa seat, and it will be a genuine good or proper use of the sofa. This means that I can relate to sofa as just a surface in this context. In a nutshell, "Surface supports Person." Replacing Sofa with Surface results in the following code:

```
Room.accommodates(Person)
Surface.supports(Person)
```

This is looking like a thing of beauty. This is Room generalised, and so are the main pieces of furniture that define each room, such as Bed, Sofa, and Chair. The two behaviours, *accommodates* and *supports*, can be coded in any programming language. This demonstrates excellent subject or object encapsulation. This code is naturally readable, understandable, and maintainable. The probability of introducing errors in the code has been reduced drastically though not eliminated yet as it also depends on the implementation of the methods. There are clearly not too many objects created in order to achieve one behaviour—one method per object. There is one level of stack trace if one wants to start gauging performance from that angle. Normally, good SOPC delivers a maximum of three levels of stack traces. This SOPC is undoubtedly reusable and composable.

The true power of SOP is in designing behaviours: algorithms. Let's consider designing the methods in SOP, having ensured that each

Subject Oriented Programming

subject (Room and Surface) has exactly one behaviour. In every Room, Person can enter and leave the room. Let's capture this by an IN/OUT flag. Since we are working on a sitting room, let's say Person enters the Room to sit. If the seats are occupied, Person stands. Therefore, the person in the room can either be standing or sitting.

Let's explore the *accommodates* method. Person enters Room. The entrance is registered. Person sits on Sofa, which is now occupied.

```
Room.accommodates(Person, IN/OUT)
Registers(Person{John}, cursor{0}, Person[John])
Gets(cursor{0}, status[SIT,STAND]): status{SIT}
Surface.supports(cursor{0}, status{SIT})
Registers(status{SIT}, cursor{0}, Capacity[SIT])
```

In each pair of curly braces is the value of the associated variable. The block brackets still represent array.

A second Person enters the room. The entrance is registered. The seat is already occupied, so he stands. Person stands on the floor. The floor qualifies as Surface as well.

```
Room.accommodates(Person, IN/OUT)
Registers(Person{James}, cursor{1}, Person[John, James])
Gets(cursor{1}, status[SIT,STAND]): status{STAND}
Surface.supports(cursor{1}, status{STAND})
Registers(status{STAND}, cursor{1}, Capacity[SIT,STAND])
```

Assuming the room is full means Person can start leaving the Room since more persons cannot enter. Notice that any of the persons can leave the *Room* in any order, but let's say the *Person* standing leaves first to simplify things.

```
Room.accommodates(Person, IN/OUT)
Registers(Person{James}, cursor{1}, Person[John,James])
|deletes(Person{James},cursor{1},Person[John])
Gets(cursor{1}, status[SIT,STAND]): status{STAND}|Same
code
Surface.supports(cursor{1}, status{STAND})|Same code
Registers(status{STAND}, cursor{1},
Capacity[SIT,STAND])|deletes(status{STAND},
cursor{1},Capacity[SIT])
```

Now, the code for Person leaving the Room has been placed alongside the entering code. The code shows that Person gets registered when entering and deleted when leaving the Room. This means that *registers* create, and *deletes* basically deletes. They both can be generalised to *updates*. The same code for Gets and *support* .means code symmetry Therefore, they are generalised to the same code. Inside Surface is another *register* alongside another *deletes*. They are similarly generalised to *updates*.

```
Room.accommodates(Person, IN/OUT)
updates(Person{James}, cursor{1}, Person[John,James],
FWD/BWD)
Gets(cursor{1}, status[SIT,STAND]): status{STAND}
Surface.supports(cursor{1}, status{STAND})
updates(status{STAND}, cursor{1}, Capacity[SIT,STAND],
FWD/BWD)
```

The generalisation of *registers* and *deletes* to *updates* needs a way to tell whether the cursor is moving backward or forward, hence the flags FWD/BWD. This essentially means the forward move is entering, and the backward move is leaving. Let's explore what *updates* really means.

```
Room.accommodates(Person, IN/OUT)
updates(Person{James}, cursor{1}, Person[John,James],
FWD/BWD)
put(Person{James}, cursor{0}, Person[John],FWD)
|put(Person{}, cursor{1}, Person[John], BWD)
Gets(cursor{1}, status[SIT,STAND]): status{STAND}
Surface.supports(cursor{1}, status{STAND})
updates(status{STAND}, cursor{1}, Capacity[SIT,STAND],
FWD/BWD)
put(status{STAND}, cursor{0},Capacity[SIT],FWD)
|put(status{}, cursor{1}, Capacity[SIT],BWD)
```

Observe that the *put*, which creates, actually puts a value in the array and then moves forward (FWD). The *put*, which deletes puts nothing in the array and then moves backward (BWD). Let's clarify what's going on in the update code really, will enable generalisation. Note that moving cursor forward means (cursor + 1) and backward means (cursor − 1):

```
IN: put(Person{James}, cursor{0}), cursor = cursor + 1 FWD
OUT: put(Person{}, cursor{1}), cursor = cursor - 1 BWD
```

These illustrate that whether creating or deleting, the cursor is always moved afterwards. This means the code for moving the cursor can be called (*move*(cursor,FWD/BWD)), and *put* remains as it is. The same thing can be achieved mathematically by adding IN and OUT as a way of generalising them.

```
put({James}), cursor = cursor + 1 FWD
+ put({}), cursor = cursor - 1 BWD
---------------------------------------------------------------
put({James}), cursor = cursor move(FWD/BWD)
```

Adding *put* + *put* = *put* (common identity), same as cursor + cursor = cursor (common identity), 1 – 1 = 0 and FWD + BWD = function that can move the cursor forward or backward as required, which is move (FWD/BWD). Therefore, the code looks like the following:

```
Room.accommodates(Person, IN/OUT)
updates(Person{James}, cursor{1}, Person[John,James],
FWD/BWD)
put(Person{James}, cursor, Person[John])
move(cursor, FWD/BWD)
Gets(cursor{1}, status[SIT,STAND]): status{STAND}
Surface.supports(cursor{1}, status{STAND})
updates(status{STAND}, cursor{1}, Capacity[SIT,STAND],
FWD/BWD)
put(status{STAND}, cursor,Capacity[SIT])
move(cursor,FWD/BWD)
```

Observe the *gets* and the *put* inside the Surface. The *gets* gets a flag—SIT or STAND—and puts it in the Capacity array. The effect of leaving can be achieved if the flag SIT or STAND is returned to the status array. But when that happens, the seat or standing spot is vacant, so it can always be replaced by another flag (VACANT). This means an introduction of a VACANT flag and pulling the *put* in the Surface to Room, near *gets*. Therefore, the status and capacity arrays are no longer needed in the *supports* method. The code looks like the following:

Subject Oriented Programming

```
Room.accommodates(Person, IN/OUT)
updates(Person{James}, cursor{1}, Person[John,James],
FWD/BWD)
put(Person{James}, cursor, Person[John])
move(cursor, FWD/BWD)
Gets(cursor{1}, status[SIT,STAND]): status{STAND}
put(status{STAND}, cursor, Capacity[SIT,VACANT])
Surface.supports(cursor{1})
updates(cursor, FWD/BWD)
move(cursor,FWD/BWD)
```

Since leaving the room is handled by *gets* and *put* in status and capacity arrays, there is no need to delete Person anymore. In fact, there is no need to keep track of the person in the room since the *gets* and *put* in status and Capacity arrays also track sitting and standing flags that clearly indicate the person is in the room. Therefore the *updates* on the person array will be removed entirely. The *updates* in the *supports* method looks exactly like *move*, so that will be removed. The following code results:

```
Room.accommodates(IN/OUT)
Gets(cursor{1}, status[SIT,STAND]): status{STAND}
put(status{STAND}, cursor, Capacity[SIT,VACANT])
Surface.supports(cursor{1})
move(cursor, FWD/BWD)
```

The code looks a lot cleaner. However, there is clearly no need for Surface support anymore, but the *move* in it can be pulled up. The Surface subject will be removed. Observe the *gets* and *put* carefully. Can you see some swapping going on? Let's set up the code to see it clearly:

```
Room.accommodates(IN/OUT)
Gets(cursor{0}, status[VACANT,STAND]): status{SIT}
put(status{SIT}, cursor{0}, Capacity[SIT,VACANT])
move(cursor{0}, FWD/BWD)
```

This makes swapping a little more obvious. Now, let's replace the *gets* and *put* with a *swap*. Let's also replace status and capacity arrays with In[] and Out[] arrays:

```
Room.accommodates(IN/OUT)
swap(In[SIT,VACANT], Out[VACANT,STAND], cursor)
move(cursor, FWD/BWD)
```

Let's then manage the invariant of the *accommodates* method so that a person cannot enter the room if the room is full and no one calls the method with an invalid argument. Let's call the array of flags that allows access to the room Services = [SIT, STAND]. The size of the Services array must always be equal to each size of In[] and Out[] for the invariant to be managed correctly outside the room. Once this Services array is in place, one can enter the room and choose to stand or sit, as long as there is vacancy, which is one of the things Services ensures. As a result of this flexibility, there is no more need for the *move*, and it will be removed.

```
Room.accommodates(Services)
Cursor = Services.pos
swap(In[SIT,VACANT], Out[VACANT,STAND], cursor)
```

or

```
Room.accommodates(Services)
swap(In[SIT,VACANT], Out[VACANT,STAND], Services.pos)
```

Subject Oriented Programming

This is truly elegant. It is readable and will surely perform well. By the way, the Services should be immutable. The use of swap in specific concise context indicates that the design is fit for the purpose. Note, that implementing the swap requires interpretation of the IN/OUT arrays and their contents.

Furthermore, the swap method needs subject for the design to be complete for the stated purpose. Naturally, as the Services *swaps* the IN and OUT array elements, the following SOPC suffices:

```
Room.accommodates(Services)
Services.swaps(In[SIT,VACANT], Out[VACANT,STAND])
```

Notice as well that Services.pos is now part of services does not need to be part of swap arguments. Interpreting the above further in SOPC in readiness for translation to Java code is as follows:

```
Room.In[SIT,VACANT]
Room.Out[VACANT,STAND]
Room.accommodates(Services):Void
    Service.swaps(In[SIT,VACANT], Out[VACANT,STAND]):Void

Service.InPosition
Service.OutPosition
Service.swaps(In[SIT,VACANT], Out[VACANT,STAND]):Void
    In[InPosition]::Temp
    Out[OutPosition]::In[InPosition]
    Temp::Out[OutPosition]
```

The following is Java code translation:

```java
public class Room {
    private InOut[] in = new InOut[]{InOut.SIT,InOut.VACANT};
    private InOut[] out = new InOut[]{InOut.VACANT,InOut.STAND};

    public Room() {
    }

    public void accommodates(Service service) {
        service.swaps(in, out);
    }
}

public enum InOut {
    SIT, STAND, VACANT;
}

public class Service {
    int inPosition;
    int outPosition;

public Service(int inPosition, int outPosition) {
    this.inPosition = inPosition;
    this.outPosition = outPosition;
}

public void swaps(InOut in, InOut out){
    InOut temp = in[inPosition];
    in[inPosition] = out[outPosition];
    out[outPosition] = temp;

    }
}
```

Designing Data Structure

Data structures are basically structures that have the capacity for the reading and writing data. The reading and writing adapter functionality of data structure should be driving the exploration design of the data structure until the eventual composable API. This is the standard SOP approach to object programming. A typical example of this is the design of the array.

Array

This is about using SOP to design behaviour itself, fundamentally an algorithm. Let's explore the idea density of the behaviour of array. Array should be able to exhibit Create Read Update Delete (CRUD). Exploring the common identities of CRUD:

Create is to update no data with new data.
Update is to update existing data with new data.
Delete is to update existing data with no data.

Therefore, the common identity is update. Exploring the common identity of read and update:

Read is to read existing data and return.
Update in the context of read is to read data and then update it.

The common identity is read, but update needs to return read data in order to generalise read at this granularity.

```
Array.updates(data):Data
```

However, read is not expressed in the method, though it is expressed in the behaviour. In order to address this, the common identity of read and update can be approached differently as follows:

> Read is to read existing data and return.
> Update in the context of read is to read data
> and then write new data.

The generalising method and behaviour is as follows, which is also an adapter, which in turn means that a second method is expected:

```
Array.readWrite(data):Data
```

The second method can be adapted to read only or write only.

```
Array.readWrite(data):Data
Array.read():Data
```

Or the following:

```
Array.readWrite(data):Data
Array.write(data)
```

Or adaptation of the behaviours' combination:

```
Array.read():Data
Array.write(data)
```

CHAPTER 5

BEYOND BASIC

SOD

As you recall from earlier chapters, the use of SOP in the design of software systems is characterised as subject-oriented design (SOD). SOD is simply the use of SOP and exploration to design for specific outcomes. Read chapter 4, and continue reading this chapter as SOD was typically used in these chapters.

In this chapter, the use of SOD and SOP is beyond basic, but things are explained as simply as possible to remove the novelty of SOP/SOD.

SOP Syntax and Semantic with Example

As mentioned previously, SOP is an elegant style of using real objects to program problems and solutions to problems for computers to compute. Real objects are physical and abstract objects. SOP uses English language grammar to express objects in a context of interest. Subject orientation emphasises the use of a simple English-language sentence to express exactly the problem of interest. A simple sentence means one subject, one verb, and at least one object. One verb explains that the sentence predicate consists of just one verb or action. Subject is an object, but a special object because of its role and position in a sentence. Subject is normally the first object in a sentence, which performs an action or actions on the second or more objects of the sentence.

Before continuing, the following example uses object-oriented language constructs in the form of class and constructor to convey some aspects of this design.

An English-language sentence is the best way of expressing the relationship between objects. The relationship between objects is what captures the problem being expressed. It is also the context of interest for the objects within which they are programmed. For example, the sentence, "House shelters Occupant" expresses House as subject, *shelters* as action, and Occupant as object. The relationship between the two objects is *shelters*, which is also the action. The sentence can be represented in SOP as

```
House.shelters(Occupant).
```

House, as the subject of the sentence, performs the action *shelters* on the object Occupant. This is the natural language of real objects expressed in an English-language sentence. Simply put, House and Occupant are the real objects being programmed to solve the problem, *shelters*.

Let's explore how the objects House and Occupant can be programmed. House is an object, but it becomes a subject when it has a dot after it as in, House.; the dot after the House makes it the subject. This implies that the action it performs should follow. In this case, the action is *shelters*. So far, House.*shelters* defines subject orientation. Action in subject orientation must be performed on at least one object. The object is not explicitly specified, which means that object(s) must be property of House or the House itself. For example, a property of House that *shelters* can be performed on is Furnishings, which is object. If House has no Furnishings as property, then *shelters* can only be performed on the House itself but as object. This means that *shelters* is performed on Room as House has at least one room. Generally, action must be directed or performed on an object at least, even if the object in context is implicit. Observing this principle enables choosing the appropriate action that matches the object(s) towards which it is directed, especially when the object(s) are not explicit.

It is a bit easier to ensure that the action matches the object it is performed on if it is explicitly specified such as:

```
House.shelters(Occupant).
```

The parentheses around Occupant signifies that the action *shelters* is a method which takes an argument Occupant. Now the code reads: House is a subject that has a method—*shelters*—which takes an argument—Occupant—and the action of sheltering is performed on the Occupant inside the method *shelters*. Inside the method *shelters* is where the two objects are programmed. The correctness of the solution programmed inside the method depends to a large extent on the accuracy of the choice of action (or verb).

Furthermore, what is the effect of programming these two objects, House and Occupant? In computing terms, effect means the output. The output of *shelters* in:

```
House.shelters(Occupant)
```

can be determined if what shelters means is explored. Let's say shelters means to shield or protect the occupant from weather. Moreover, a house shelters somebody if the person is in the house. The person who is sheltered is an occupant. This means that Person should be the argument of the *shelters* method and Occupant the output of the method.

```
House.shelters(Person):Occupant.
```

The colon after the closing parenthesis indicates that

```
House.shelters(Person)
```

is input, and the Person who follows the colon is the output.

Subject Oriented Programming

Let's explore what happens when an object instead of an action comes after the dot. For example, House.Room means that the subject House has an object, Room, as a property. There are two types of subject property: subject intrinsic property and subject inheritable property. In this example, if a given House is a one-room house, then Room is an inheritable property for House, which means that House inherits the characteristic or behaviour of a Room. But if the given house has more than one room, then Room is an intrinsic property of the House.

Let's explore the interaction of House and Person. They interact via action or method: *shelters*. In this context, there are two subjects interacting, the House and the caller of the method *shelters*. The caller interacts with House by inserting the argument Person and then calls the method *shelters*. These two interactions allow for lazy evaluation of the method *shelters*. For example, the caller can insert the argument Person and refrain from calling *shelters* just until it has inserted other arguments or prepared other computations. Then it can call the methods in parallel or serially as it chooses.

Once the method *shelters* is called, the interactions are now between subject (House) and object (Person). Typically, the House interacts with the Person by calling a method on Person. The effect or output of this call will be used to interact with the intrinsic properties of House. Overall, there are two types of interactions in a method. This means that more than these two interactions in a method is unnecessary and should normally be avoided. One of the ways of avoiding unnecessary interactions in method calling is to ensure that only one method call is enough to achieve a purpose. This helps to avoid layers of method calls. For example, the method call on the Person argument must not dispatch a call to another subject.

Let's explore the concept of interface as the surface of interaction between two subjects, House and Client. Normally, interface defines the function for polymorphic behaviours of objects. Let's say Person is an interface between subject (House) and subject (caller or client) in

```
House.shelters(Person):Occupant.
```

Person is an interface with a function in the form of method, but the name of the method or action is not known yet. Before mentioning the polymorphic objects that conform to the single behaviour of the interface, let's name the method or behaviour for object. As soon as the Person is inserted and *shelters* is called, the next interaction is for *shelters* to act on Person. Let's explore what it means for a House to shelter Person. When a house shelters, it prevents rain from falling on a person or prevents the person from getting wet. But the person can still get wet by some other means in the house. In the same vein, a house prevents direct sunlight from hitting the person along with the associated heat, but the person can still get hot in the house. Furthermore, a house prevents adverse cold from touching the person when inside the house, but the person can still feel some cold inside the house. Similarly, a house prevents creepy things from entering the house, but every now and again, some of them do get into the house. All these mean that the sheltering the house provides has a level of weather control inside the house. Hence the method name,

Subject Oriented Programming

```
Person.controls(Weather).
```

Furthermore, let's quantify rain not touching the person inside a house. Rain is basically water in the air. It is the amount of water in the air that determines if it can be categorised as rain or on the other end of the spectrum, dryness. Therefore, it can be captured with humidity, which is the quantity of water in the air. For the hotness or coldness of the air, temperature can capture them. For the number of creepy things or pests in the house, they can be captured by pest population or using idea density, pestulation (new term). These mean that Weather can be changed to condition. The reason for idea density of a house that results in this code will be clear shortly.

```
Person.controls(Condition)
```

Condition in the house is an object that consists of three properties: humidity, temperature, and pestulation. But naturally, it has no behaviour. However, Condition in the house can be simplified to Weather and Pestulation, where weather comprises humidity and temperature. As a result, the effect or output of sheltering is the condition in the house which consists of Weather and Pestulation. The following SOPC illustrates these:

```
House.shelters(Person): Condition
Person.controls(Weather, Pestulation)
Condition(Weather, Pestulation)
```

Note that there is always an effect or output for every behaviour or method, but sometimes it may not be obvious what that output really is. Returning the output of a method for further use or reuse can be optional depending on need. On this occasion, the Condition of the house has been returned from the *shelters* method instead of the Occupant.

One of the reasons for returning object(s) from a method is so that they can be used by another method. Therefore, returning object from a method is one of the ways to compose computation or behaviours and reuse their functionality. Let's say that the Condition returned from the *shelters* is needed as an input another action *controls*. How can it be expressed in SOP?

```
House.shelters(Person):Visitor.controls(Condition):Void
```

The void on the end of this SOPC means that no object was returned and marks the end of computation chain. In SOP, more than one value (or object) can be returned from a method if the values or objects are not easily generalised and not enough be put in an array. For example, *shelter* could be returning two objects (Weather and Pestulation), which will be passed forward to *controls*.

```
House.shelters(Person): Weather, Pestulation.
```

When composing, House and Visitor will be as follows:

```
House.shelters(Person):   Visitor.controls(Weather, Pestulation):Void
```

If the outputs of two methods are the two inputs of another method, they can be composed as follows:

```
House.shelters1(Person),House.shelters2(Person): House.shelters(Condition1, Condition2): Condition
```

Subject Oriented Programming

If a method outputs two values (Value1 and Value2) and another method needs only one of the values as input,

```
House.shelters1(Person): Value1, Value2.
House.shelters1(Person): House.shelters(Value1):Void
```

Notice that the essence is ensuring the composability of API at every level of abstraction that addresses the problem being solved. For some use cases or problems, this abstract level could be the optimal API design. Therefore, the composition of the fluent API may be ideal and different from fluent APIs in Java, for example. Hence, as similar as SOP is to actual object code in existing programming languages, this is still design and not actual program code. The ideal nature of SOP syntax allows the design exploration to get to the pure, elegant, complete solution as shown below.

If a method takes two inputs, and another method supplies only one of the inputs. And the other input comes from another source, such as the property of the enclosing class, they can be composed as follows:

```
Property Condition2
House.shelters1(Person):House.shelters(Condition1,
Condition2): Void
```

Let's say that the House will keep track of the persons who are sheltered. This means that house needs a property to keep record of the number of people in the house or sheltered. The shelters method will look like the following:

```
House.Condition
House.shelters(Person): Condition
```

```
Person.controls(Condition)
personCount++
```

As shown in the code snippet above, House interacts with the Person (external interaction) and then with personCount property (internal interaction). That's two interactions so far, but let's see the inside of the Person.*controls* method.

```
Person.Condition
Person.controls(Condition): Void
Condition.Temperature,Humidity,Pestulation  ::  this.
Condition.Temperature,Humidity,Pestulation  //1

Condition.*:this#condition.*  //2

Condition.*=this#Condition.*  //3

Condition.Temperature
Condition.Humidity
Condition.Pestulation
```

The above code shows three ways of setting property values of the subject (Person) concurrently. The first way is the pure SOP way of concurrently setting or constructing an object or subject. The other two ways are typical procedural ways of constructing objects or subjects. Since the setting of the properties is done in parallel, it means the number of interactions is just one—an internal interaction. This makes the total number of interactions between House and Person three.

Let's explore *shelters* in the context of house condition (temperature, humidity, and pestulation). In light of the subject orientation Condition, object is a sensor for temperature, humidity,

Subject Oriented Programming

and pestulation. Since Condition is a sensor, it needs to be in the House to sense its interior condition. It is by way of this sensor that the house expresses *shelters*. But how does the Person receive this action from House? Person receives the action by feeling the condition in the House. Person feels the temperature, for example, by reading the temperature off the sensor (Condition or Services). Reading is the operative term for reading computer memory, but in SOP terms, it is more like having access to temperature value in the enclosing subject.

First, let's unpack the meaning of the SOP reading and then reconcile it with memory reading of values. The Person object that is passed into the shelters method immediately has access to Condition object, which in computer memory terms is just a memory address of the subject (Condition) and not the address of the temperature. This means that the Person does not feel or read the house temperature directly in the house, which is not the solution of interest here. The solution of interest is when the Person enters or is sheltered, he or she should feel the effect of sheltering directly.

Therefore, there is a need to unwrap Temperature, Humidity, and Pestulation from Condition and put them in the House as follows:

```
Temperature
Humidity
Pestulation
House.shelters(Person): Services
Person.controls(Temperature,Humidity,Pestulation)
personCount++
```

This piece of code gives the impression that the Person object immediately feels or can read the temperature in the house directly. In computer memory terms, the memory address of the temperature value is directly at the disposal of the Person object. The zero copy principle in computing means that data should only be read for computation and not transferred from one memory location to another. So, let's look at what this means in SOP terms. Let's say that the House temperature is 25 degrees Celsius, and the body temperature of the Person who enters the House is 35 degrees Celsius. In reality, the House or room temperature does not just get added to the body temperature. Otherwise, the body will basically overheat. Let's say that 25 degrees Celsius is enough to raise the body temperature by 2 degrees Celsius, which means that the body temperature of the Person is now 37 degrees Celsius. Temperature is modelled in same way as speed and can be scaled in orders of magnitude smaller and bigger. Humidity, on the other hand, is the quantity of water in the air in a given environment. Humidity is measured from 0 (total dryness) to 100 per cent water in the air in a given environment. Pestulation is the number of pests in a given environment. Pestulation is measured from zero to the number of pests that can be contained in the given environment—population density of pests. The common identity of the three objects is that they are measured from zero to a certain number or quantity. But temperature can be scaled to smaller or bigger spans, which means different levels of zeros on lower-bound edge cases and as well upper-bound edge cases. Pestulation can be scaled in similar way as Temperature, albeit in increasing orders of magnitude. Humidity is scaled from 0 to 100 per cent, but 100 per cent means ratio of number against cent. Basically, 100 per cent, which can be represented as 100:1cent. This generalisation is to ensure design completion, and nothing more really.

On the other hand, another type of method, constructor, is the same typical method in argument insertion and method calling, but it does a bit better in interactions. For example, if shelters were a constructor, Person would interact only with the intrinsic property of House. Therefore, there is only one type of interaction in a constructor.

It is important to note that the sentence House shelters is a valid sentence that expresses subject orientation. This generally orients how the relationship shelters between the subject and objects should be programmed. Having said that, let's generalise the above SOPC to the following for translation to Java code.

```
House.Condition
House.shelters(Person):Service
    Person.controls(Condition):Void
    House.In, House.Out:Service

Person.Condition
Person.controls(Condition): Void
    Condition.merge(this.Condition) ::this.Condition
```

The following is Java code translation of these SOPC.

```
public class House {
    private final InOut[] in = new InOut[]{InOut.
    VACANT,InOut.SIT};
    private final InOut[] out = new InOut[]{InOut.
    STAND,InOut.VACANT};
    private final Condition condition;
```

```java
    public House(Condition condition) {
        this.condition = condition;
    }

    public Service shelters(Person person) {
        person.controls(condition);
        return new Service(in, out);
    }

}

public enum InOut {
    SIT, STAND, VACANT;
}

public class Person {
    private Condition condition;

public Person(Condition condition) {
    this.condition = condition;
}

public void controls(Condition condition) {
    this.condition = condition.merge(this.condition);
    }
}
```

Designing Algorithm

In SOP, algorithm corresponds with the verb, action, message, method, and behaviour. The design can start with the verb and then subsequently, the Subject and the Object of the sentence can be discovered with exploration. This is modelling algorithm only as this is not an algorithm book. However, this is how to explore the design to a level of abstraction or detail that supports clarity in the implementation of the algorithm. The following sorting algorithm exemplifies SOP/SOPD.

Sorting

Sorting is simply putting a set of things in known order or ordering a set of things to a known sequence or series. Sorting is chosen as a case study because it is a popular area of computing and algorithm-oriented, which is very important in showing the use of SOD.

Let's start with idea density of sorting and array of numbers, represented in SOP as,

```
sort(Array[5,9,1,6,3])
```

So far, there is no subject in the SOP code above. We will get to it in the course of the exploration. It is an action and object-sort array.

Conventionally selecting an element in the array as a pivot (smallest, biggest, or average) to compare other elements against it places each element in the proper place in a given order. This sorting method, or indeed other methods, involves determining the place of each element. Therefore, sorting, meaning, and determining the index of each element of a given array, whatever order, is required. This means that the common identity for sorting is the index. Therefore, determining or identifying indexes of elements of array of the set of things to sort determines the algorithm. There are many algorithms for sorting: quick sort, merge sort, selection sort, insertion sort, and so on.

The idea density of sorting has identified the common identity of sorting. Let's represent its subject-oriented code as:

    Indexes.sort(Array[5,9,1,6,3])

Continuing the exploration of the idea density of sorting, the indexes of the array being sorted are the common identities for each element, which means they are the same for every element. This implies no difference for each element of the array. Therefore, in the context of indexes, there is nothing to sort as the elements are already sorted by default and insertion order. This feels strange as the resulting array is not in the expected order—serial order. This is the type of clarity SOD provides.

    Indexes.sort(Array[5,9,1,6,3]): [5,9,1,6,3]

The expectation is to sort the array in a different order. Therefore, another property of the elements that provides differences must be used. This means the nature of the elements must be known in order to sort. But if the natural property of the elements of the array is not apparent, then insertion sort becomes the sort even if it is not the order required.

Let's say the array of numbers above are of oranges, and the diameters of the oranges are being considered. One could argue, "What differences do the diameters bring that the numbers themselves in the array couldn't have done?" The difference is represented in SOPC as follows:

    Diameter.sort(Array[5,9,1,6,3]): [1,3,5,6,9]

Notice that indexes are an inherent property of array and subsumed in the array, which allows diameter to be the subject as the array is object in the code. The indexes are used as contextual common identity, where diameters are the common inequality that can be sorted.

Continuing the exploration with generalisation of the previous code, in SOP, generalisation has been both specific and general. This ensures modularity and composability of the resulting API. For example, the diameter in the SOPC above needs to be generalised in a specific way. Instead of diameter as the subject, it could be colours or types of oranges that are there as common inequalities. In the same vein, it could be mangos or apples instead of oranges. At this point, the idea density of the subject requiring specific generalisation can be felt. The properties—diameter, colour, types, and so on—being sorted can be categorised as order. Something along this line that is both specific and general will do, but order fits the bill here. However, order has different connotations in domains like finance, e-commerce, and so on from what is portrayed here. Therefore, sequence is used instead of order for clarity. The code is generalised as,

```
Sequence.sort(Array[5,9,1,6,3]): [1,3,5,6,9]
```

This raises an immediate question: Does order know how to sort all things in the computer system? What about the specific generalisation on the object to ensure that? One can reasonably expect that properties being sorted can be represented with numbers, alphabets, alphanumerics, or strings. There can be a layer of view associated with properties that gives them further meaning from what's obtainable in strings, for example. Java programming language expressed this in equals method. All in all, it culminates in string representation of the properties being sorted.

The subject sequence provides the sort algorithm and implementation details that handle all of these properties. Suffice it to say, there is no need to do more about the implementation details of the API.

Summary

SOP basically provides a framework and syntax for object programming that captures the confluence between imperative programming and functional programming as illustrated fundamentally throughout this book. SOP roots object programming to English language grammar and lays the foundation for future automation of object programming with SOP dimension or core syntax. With SOP dimension framework, high quality object programming is streamlined to be a bit more straight forward. Furthermore, all of these are geared towards object programming automation. Realistically, the automation of object programming may not be in the distant future with the pace of advances in artificial intelligence (AI) and the desire for AI to design and write subject/object oriented programming code or application software in general.

Milton Keynes UK
Ingram Content Group UK Ltd.
UKHW021334290324
440221UK00002B/13

9 798823 086714